DEATH COMES TO THE NURSERY

Books by Catherine Lloyd

DEATH COMES TO THE VILLAGE

DEATH COMES TO LONDON

DEATH COMES TO KURLAND HALL

DEATH COMES TO THE FAIR

DEATH COMES TO THE SCHOOL

DEATH COMES TO BATH

DEATH COMES TO THE NURSERY

Published by Kensington Publishing Corporation